Dorman B. Eaton

Metropolitan Health Bill

Dorman B. Eaton

Metropolitan Health Bill

ISBN/EAN: 9783337155889

Printed in Europe, USA, Canada, Australia, Japan

Cover: Foto ©berggeist007 / pixelio.de

More available books at **www.hansebooks.com**

METROPOLITAN HEALTH BILL.

REMARKS

OF

D. B. EATON, Esq.,

AT A

JOINT MEETING OF THE COMMITTEES

OF THE

SENATE AND ASSEMBLY,

ALBANY, FEB. 2d, 1865,

WITH

AN APPENDIX.

Published by the Friends of the Bill.

NEW-YORK:
GEORGE F. NESBITT & CO., PRINTERS AND STATIONERS,
CORNER OF PEARL AND PINE STREETS,

1865.

METROPOLITAN HEALTH BILL.

Mr. Chairman,—

I have been requested to appear before this Committee, and to state some of the reasons why a Metropolitan Health Law is demanded by the city of New-York. It will readily occur to every reflecting person that laws properly adopted to promote public health must always be, in some respects, peculiar, as they are equally demanded by the most disinterested principle of humanity and benevolence and by the lowest instincts of self-preservation and fear. The meanest coward, and the noblest philanthropist, by different motives, are prompted to common action.

Until a considerable advance in civilization there is scarcely a possibility that adequate health laws should be enacted, and the character of such laws and the efficiency of their execution will be found to keep pace with, and illustrate the disinterested sanitary science, and the intelligence of every nation. Such laws may also bear some relation to the realization of danger to public health.

It requires the existence of a high degree of scientific cultivation and varied experience relative to the causes that produce disease in all its manifold forms, before any people will properly appreciate either the want or the utility of proper health regulations. There are few causes more varied and profound than those that relate to the preservation of public health and the prevention of disease.

Many of these causes are local, and hence a knowledge of soil, climate, habits, buildings, and drainage, are involved. And hence the health regulations of each city must be adapted to the peculiar dangers and exposures of that city.

---

* In the Appendix will be found an Abstract of the pending Bill, and a list of the present Health Officers and of their salaries.

The dangers to health inercase vastly with the population of plaees, and the density of such population. Density
of populations endangers health in two ways: first, by impair-
ing ventilation and otherwise producing habits and eonditions
that *originate* disease; and secondly, by greatly inereasing its
spread by exposure and contagion. Such results also eause
mueh more study of the sourees of danger to pubiie health
in large cities, and demand more stringent measures for their
removal.

For this reason, laws relating to public health, have
usually originated in large cities; and it is in the older coun-
tries, and in the experience of the great eities, that examples
of adequate health laws are to be sought.

It is only within a few years that there has been very useful
legislation on the subject of public health in the United
States, and the few illustrations found relate mainly to New-
York, Philadelphia and Boston. In this country, therefore,
this field of legislation is yet to a great extent unoccupied.

It is impossible that a health law should deelare mueh
more than authority in general terms, leaving the officers that
administer them to apply them according to the exigeney of
the danger. Moreover, it is apparent that the execution of
a health law requires great intelligenee and experience in the
offieers. The duty of the Health Officers is to study the
causes of danger to publie health and diseover their remedy,
and then to apply the power he possesses to the removal of
the causes. This implies a sort of *minor legislation* all the
time on the part of the Health Officer. He is not simply di-
reeted to do *a specific thing*, but is required to *discover hidden
causes, to devise the proper remedy*, and then to exereise the
legal power of removal and redress.

It is not the provinee of a Board of Health to simply
cure diseases or consult the health of individuals. Sueh
boards are to attend to the pervading dangers that affect the
general publie health. To devise and apply wise precautions against the hidden causes of disease which threaten the

health of communities or classes, is their especial duties. These precautions relative to the soil we dwell upon—the houses we live in—the drainage of waters—the food and drink we consume—the places of public gatherings—the filth that poisons the air—the trades that contaminate localities—the decoctions and drugs that are used as remedies—the precautions against contagions—the arrest of pestilential disease.

A law of 1850 affords a striking illustration of a health law, as applied to the draining of a swamp.

(See laws 1850 p. 585, ch. 271.)

It must be apparent that such duties can be performed only by scientific and intelligent officers who act in concert, and for a series of years upon a wise systematic method. Many of the most important results sought can only be reached by a long and careful study and by comparison of systematic observation and experiments.

It would seem very plain that a Board or Commission with continuity of succession that should make the subject of public health a special topic of study and observation, would be best calculated to attain the end sought; and the science of health must from its nature be a progressive study, and all health boards should lay before the community full reports of their experience and observation as well as of their ministerial acts.

One of the earliest health laws, marked by any noticeable comprehension, is the English law of 1848, relating mainly to London. That law originated in a " Report of her Majesty's Commissioners of Health." In 1847 it was formally announced in the English Parliament, that " Her Majesty had thought fit to appoint a commission to report on the best means to improve the Health of the Metropolis," &c.

That Report resulted in the elaborate law of 1848, which created a "*Metropolitan Board of Health*," and gave it large powers.

In 1858 a more elaborate Act of Parliament enlarged the health powers of this Board, and extended the law to the other larger cities of the United Kingdom; and it is well known to all who have given the subject of public health the attention it demands, that the effect of these laws has been to lower the death rate, to lessen disease in the great cities of Great Britain, and to extend the term of human life. This Board has been an intelligent centre of benevolent, sanitary and scientific inquiry, and has enlisted, to a very salutary extent, public interest in the sanitary condition of the lower classes and aroused a concert of action for the promotion of public health throughout the United Kingdom. No corresponding action has yet been taken on this side of the ocean, but in France a similar work is in progress.

In New-York a similar work is imperatively demanded. It has every natural facility for being a very healthy city, but is more unhealthy than London, Paris, Philadelphia or Boston. The main cause of this unfortunate result is believed to be found in the negect of those precautions which wise health provisions, wisely administered, would, in a few years, afford.

There is not, and has not been, in New-York, any body deserving the name of a Board of Health—any body that embraces any considerable portion or representation of the educated mind, of the sanitary experience, of the medical science, of the benevolent heart, or even of the practical administrative talent of the city.

The organizations that have had the exercise of health powers in New-York, have been mainly political bodies, or at best, bodies that devoted most of their attention to political and general municipal affairs. The subordinate officers have, in the main, been utterly unqualified to act as health officers, and have had a variety of conflicting duties to perform, and have been appointed largely in reference to political opinions and influences.*

Besides the health powers are distributed in such a manner as to cause great uncertainty as to who is authorized

---

* See Extracts from Mr. Delevan's Report in the Appendix.

to exercise them; and there is a necessity of consulting so many officers, that concert of action is nearly impossible, even if the officers were capable and desirous of serving, with singleness of purpose, the cause of public health. *The result has unquestionably been that the Health Laws applicable to the city of New-York, have never been usefully or wisely administered; and, as now delegated, they are found to be nearly impracticable; the public health is greatly and needlessly exposed, and by reason of the disease spread from the army, is now much more endangered than ever before.*

Disease cannot be confined to localities where it originates or most readily spreads, but is soon carried along the railroads and rivers, and on the wings of the wind, and speedily finds its way into all parts of the State and nation. Hence, while the people of the city are directly interested in its sanitary condition, the danger in the city is yet the common danger of all the people of the State.

The existing organization and the principal authority concerning public health in the city of New-York, are derived from the 275th Chapter of the Laws of 1850. (Laws 1850, p. 597.)

It will require but little examination to see that the law is so imperfect that its efficient execution is nearly impracticable, and the strange provisions almost force one to the belief that there was a design in their confusion.

§1. Confers all *legislative* power on the subject of health upon the Mayor and Common Council. It is not easy to understand the force of the words "*legislative* power."

§2. The Mayor and Common Council, when acting in relation to health, are made a *Board of Health.* Ten members make a quorum, and the Mayor is made President of the Board.

Now, it is known there are two Boards of the Common Council, and it is not said whether all the ten composing a Board of Health may belong to one Board, or must belong in part to both, to make a legal Board of Health. Besides, the two Boards of the Common Council are composed of more than twenty members, and, *hence, there may be two separate legal Boards of Health in existence at the same time; and what ten politicians ordered to-day as a Board of Health, twelve or fifteen other members of the Common Council,* (who did not act to-day,) *may reverse to-morrow,* and, as a separate Board, may make other orders.

Besides, the Board of Health has no power, save when the Mayor chooses to call the Board together, and, as often has been the case, the Mayor will not make such call. The law does not provide whether he need *call* more than ten members to make a legal Board. There has been no meeting of the Board, it is believed, since 1863.

The practical effect of this power of *calling* the Board together by the Mayor is well illustrated in the Report of the City Inspector. (Mr. Boole, who exercises at this time the principal health powers in the city,) who says, in his Annual Report, filed in January, 1865 :

"Neither my desire, nor that of the Common Council, has been able to induce the Mayor to convene the Board of Health, and the city is thus deprived of their action on matters of pressing importance, when their authority is *absolutely necessary for efficient action.*"

Probably many good citizens agree with the Mayor, that the only action likely to be taken would be quite foreign to public health, and would be very dangerous to the public purse. So, in the meantime, and at this threatening period, *there is almost no execution of the health laws.*

§3. This provides for *secret* sessions of the Health Board.

§4. Creates a *distinct* Board, known as "Commissioners of Health," and

§5. Declares the duty of the latter to be, to "*advise* the Mayor and Board of Health," and also the "City Inspector." It will be observed that *no physician* is a member of the Board of Health, or has any authority to be present or ask a hearing at its meetings.

§6. A Health Officer is created, and he is to perform the duties which the Mayor and Commissioners of Health OR the Board of Health shall order. The orders of such separate bodies may be quite inconsistent, and who the City Health Officer shall obey, is quite uncertain. The result is, he does nothing, or does as he pleases.

§7. This Health Officer may appoint an assistant who may perform *all his duties*. Hence, the principal may do nothing, and his office be a sinecure, and the duties all be devolved on his assistant, who is *not* named or removable by, or responsible to, *either the Board of Health or the Commissioners of Health*.

§8. Defines the duties of the Resident Physician, and makes it the duty of this doctor to attend to the private wants of any person who is "reported" (by whom?) to either the Board of Health, to the Mayor, or to the Commissioners of Health. It is plain the Resident Physician is clothed with no duty properly relating to public health. No class or description of persons are directed to be attended to by that officer.

§9. The Health Commissioner is created by this section. He is merely an assistant of the Resident Physician.

§10. This requires that the Mayor and Commissioners of Health shall meet daily, but as they are only to *advise* the Board of Health, *which does not meet at all*, and in regard to which there is no provision for meeting, this section is quite remarkable, even as compared with the other parts of this remarkable law.

§11. This gives the Resident Physician a salary of $1,250; and the Health Commissioners an *annual salary of $3,500*. Now, as the Resident Physician is, by the fourth section, made one of the Health Commissioners, he draws a salary in each capacity, and is, therefore, entitled annually to $4,750.

The sums paid to these several officers, for pretended services as Health Officers, are, therefore, in the aggregate, not less than $22,250 annually; and it may be said, without ex-

2

aggeration, that they perform no valuable services in the sphere of their public duties, and only the two physicians perform even a limited service of any kind.

The Health Physician and Health Commissioners are appointed by the Mayor.

§12. This section authorizes the Board of Health to appoint an unlimited number of persons to office.

Title 2d, (extending from pages 599 to 607,) of this Health Law of 1850, confers upon these impracticable health tribunals a large portion of all the authority conferred by law for quarantine purposes. The tribunal was so inefficient and useless, that the exigencies of quarantine service caused that power to be taken from the Health Board altogether. And finally, by a law of 1863, (Laws of 1863, p. 573, ch. 358,) a regular Board of Quarantine Commissioners was established, and the power over quarantine was given to that Board, and it is not proposed to interfere with that power or Board.

Similar relief is now greatly needed for the city of New-York.

Returning again to the Health Law of 1850, p. 607, as applicable to the city of New-York, we find Title 3d, Article 1, §1, provisions that greatly add to the confusion and uncertainty before referred to. For large powers over public health are conferred specifically on the "City Inspector of the City of New-York." Whether these are a part of the *legislative* powers before mentioned as conferred on the Mayor and Common Council, or are a different class of powers, is a matter of great perplexity and doubt.

Certainly the powers of the Inspector are very extensive and unguarded.

(1.) He may appoint, with the consent of the Aldermen, "so many Health Wardens and *other officers*" as "the Common Council *or* Board of Health shall direct," &c., "such Health Officers and Wardens, &c., are subject to the *control of the City Inspector of the City of New-York.*"

We have seen that the two Boards of the Common Council, (with the Mayor acting as President,) or *any ten of them being a quorum*, constitute the Board of Health. Hence *any such ten members*, on the nomination of the Inspector, may appoint just such Wardens and officers as they see fit.

(2.) These Wardens and officers are to report the sanitary condition of the city, not to any Board of Health, but to the Mayor and Commissioners of Health. We have seen that the duty of the latter officers or Board is only to *advise* the Board of Health. Was there ever such confused, impracticable and ill devised legal machinery. There is no provision that any of these Wardens or officers shall be physicians or men of science or experience. It is well known they are, most of them, mere politicians who draw their large salaries and otherwise utterly neglect their official duties.

(3.) Then follows sundry other powers given to the City Inspector, including cleaning of lots and places, but he is obliged to consult the Mayor and Commissioners of Health as to such matters.

(4.) Then, at page 608, § 2 & 3, "the Mayor, Aldermen and Commonalty of the City of New-York," &c., are authorized to make all By-laws they may deem necessary to preserve the public health.

It will be seen that the exercise of these powers is quite inconsistent with powers given to the City Inspector alone, and it is no matter of wonder that the health regulations of the city are in a state of inextricable confusion.

Whether the body referred to in § 2 & 3, page 608, is the same body as that referred to in § 5, page 609, and in § 2, page 597, is by no means easy to determine.

If the various powers conferred on the Board of Health, Mayor and Common Council, Commissioners of Health, and City Inspector, at pages 607 to 615, were all conferred upon a single Board of Health, made up of six or eight intelligent persons of whom one-half were experienced physicians, these powers could be exercised with a third part of the cost to the city, and in that wise, harmonious and efficient manner that would avoid the present confusion, and vastly promote the health of the city of New-York.

Sections 10 to 13, inclusive of pages 610 and 611, (Laws, 1850,) afford striking illustrations of the confusion that follows, conferring power on so many different officers with such undefined relations. It is not a matter of surprise the reports required are almost never made.

As no small portion of all the health powers really exercised in the city of New-York, are that portion conferred upon the City Inspector, and it is that officer that has expended those vast sums of money that have raised so much alarm, it is proper to consider the origin and nature of his department and to call attention to its administration.

The amendments to the City Charter, (Laws 1849, ch. 187, p. 278 to 285,) established numerous departments in the City Government. The 16th section of the Act creates a department in the City Government to be called the " City Inspector's Department," and it is declared this department " *shall* " have cognizance " *of all matters relative to the public health* of " said City, and the chief officer thereof shall be called the " City Inspector." This looks more like a concurrent than like an exclusive jurisdiction.

The 20th section provides that all Heads of Departments shall be elected every three years by the people.

The term was changed to two years by the 21st section of the Act of June 14th, 1857, ch. 446, and by the 19th section of the same Act, the Mayor appoints and the Aldermen confirm the City Inspector.

And by the 27th section of the last cited law the power of the City Inspector's Department was further modified and is now as follows: " The City Inspector shall *have cognizance* " of all matters affecting the public health, *pursuant to the* " *ordinances of the Common Council* and *the lawful* require- " ments *of the Commissioners of Health, and of the Board of* " *Health*. There shall be a bureau in the City Inspector's

"Department to be called the Bureau of Sanitary Inspection "and Street Cleaning, under the control of an officer named "the Superintendent of Sanitary Inspection," etc., etc., and that officer is " to remove nuisances detrimental to public health," etc.

See Valentine's Laws relative to city of New-York p. 275.

When the powers given the Inspector in 1849, and those given in 1857 are compared with the before-recited law of 1850, it will appear to be almost impossible to decide what power relative to the public health are at this time in the City Inspector; what in the Board of Health ; what in the Mayor and Commissioners of Health ; and there is no occasion for surprise—that the City Inspector is the only officer that enters upon the exercise of any health power. Yet this officer is, at this moment, under sentence of removal or suspension by the Mayor, under 21st section of the Charter of April 14th, 1857 !

Hence it is, *that at this time, the vital and responsible powers over the public health, (distributed with such confusion through so many bodies,) and the vast patronage of street cleaning and general inspection in the city, and the annual expenditure of a million of money, are all now in the control of a single officer, who is at the head of one of the numerous subordinate departments of the City Government. His power and patronage are vastly greater, at this moment, than those of the Mayor, Aldermen and Councilmen of the city united together. He employs more men, dispenses more money and exercises powers more vital to the public well-being.*

And it cannot be denied by any one who will look into the facts, that a large portion of the shameful abuses to which this condition of things has given rise, results naturally from the confused and impracticable laws under which the powers of the City Government *are* exercised.

No extended reference can here be made to the Ordinances of 1859, under which the immense powers of the City Inspector's Department have been consolidated and made effectual for those vast expenditures and flagrant abuses which now alarm and arouse the public mind. They give the most important powers of municipal legislation in the city of New-York. The ordinances of this department alone, as published in 1860, fill ninety printed pages, and no one can read them and any longer wonder at existing abuses.

A copy of the Report of F. I. A. Boole, City Inspector, for the year 1863, dated July 22, 1864, is before me, and was published at the expense of the city. It fills 518 closely printed gilt-edge pages, is bound in *morocco*, and is beautifully *embossed and adorned with gilt letters and designs* and the name of the donee, in gilt letters, appears on the gorgeous cover. He was doubtless a *valuable friend* of the Inspector. Any one who has seen this rich specimen of official bookmaking will ever after read a corporation printer's bill with diminished surprise, and loose half his wonder at the millions paid for city inspection and street cleaning. Not much can be said here of its significant contents.

There were, in 1863, in the city of New-York, of marriages, 3,272; of births, 6,426; of deaths, 25,196—as Mr. Boole reports, at page 7. At page 10, he makes certain comparisons of the death rates of London with those of New-York, so imperfectly as to be no guide, it is true. Yet he makes the figures show that the rate of deaths is higher in London than in New-York, and then triumphantly calls upon the community to give the credit "due to the *enforcement of sanitary* regulations in New-York," which have, he thinks, contributed to the result he reports.

There was never a more unfortunate request. He assumes, in his comparison with London, the correctness of his own returns, and asks to be judged as to his manner of enforcing sanitary regulations on the basis of the correctness of those returns of his Report. Now, by these returns, in 1863, 6,426 human beings were born, and 25,196 died, in the city of New-York, a *net decrease of population*, and a total loss, under Mr. Boole's enforced sanitary regulations, of 18,770

human souls in a single year! Perhaps, if this be a fair specimen, false reasoning, presumption and extravagance were
never united in larger proportions than in the City Inspector
of the City of New-York.

It is hardly to be doubted that the returns of marriages are
very defective, and it is manifest that those of births or deaths,
and, perhaps, both of them, have not an approximation to
truth.

Yet this officer, in his gorgeous Report, at page 11, says,
"It is in vain that those who have not access to *reliable statistics*," (of course, meaning the Reports of the aforesaid City Inspector,) pretend that New-York is not the most healthy of
the great cities of the world.

It is not the purpose of these remarks to establish the
solemn truth, that the contrary is the fact. Nor can I stop
to expose the other and frequent errors and assumptions of
this extraordinary Report.

There is a law, of the 2d day of April, 1853, relative to
births, deaths and marriages, and providing that the City Inspector of the City of New-York shall keep a record thereof,
and that there shall be collected a fee for the registration of
the same, of such amount as the city authorities may authorize. It may be suggested that it would have been interesting
to the public to have known the rates charged for this registration, the number of thousands of dollars collected by the
City Inspector for such service and the disposition made
thereof; we find nothing on the subject in this gilt-edged
Report, or in the Report of the Comptroller. Where is the
money?

With the unparalleled present powers of the City Inspector's Department, it is a little surprising to read at p. 12, that
the City Inspector has repeatedly demanded, in vain, *additional* powers "to render the powers of the City Inspector
more effective"—for what?

And it is some proof of the impracticable nature of the
laws we have referred to, and illustrates the fatal confusion

and hostile rivalry that grows out of the present delegation of the power over public health, to read this language at the 12th page of this gilt-edged Report. "It is *not true* as alleged "by his Honor the Mayor, in his late communication to your "honorable body that I possess a power that no *honorable* "*man should desire to exercise.*"

Does not a solemn sense of public duty to a great city, whose miserable government, reeking filth and pestilential air, not only degrade the morals and endanger the lives of its own citizens, but those of the people of all the State, sternly demand of the Legislature some prompt and efficient remedy for those unseemly official conflicts and criminations—those confused and impracticable laws, and those alarming consequences ?

We agree with the declaration of Mr. Boole's Report, at p. 13, that in London and Paris "it is a fact worthy of all praise," &c., "that all that concerns public health receives the prompt attention of government," &c., and, we may add, the appropriate attention of metropolitan health bodies. And we now ask that the Government of the State of New-York will attend to the same subject and establish a similar Metropolitan Health Board to those of London and Paris. We further agree with the City Inspector, again at p. 13, where he says that the "present system *is deficient*," and that "at-present the City Inspector is obliged (?) to *assume* responsibilities *unsanctioned by municipal ordinances*, but imperiously demanded by local exigencies."

At page 21, it is stated that, in the city of New-York, 6,000 families live in underground cellars, which nurseries of disease are inhabited by 18,000 persons. He says his Health Wardens *daily report* the state of all those localities. It is to be *hoped they do*, and that they are good, devoted men, who make these reports, and that none of them are mere political hacks and keepers of groggeries; but the contrary is believed, and no reports have ever been seen.

"It ought to be stated that the Report proper, containing

the result of all the City Inspector's reflections and experience for the public instruction in 1863, only fills 24 printed
pages, and the great bulk of the volume is filled with extracts
from foreign work which only illustrate how far this city, in
regard to its health regulations, is behind the other great cities
of the world, and what a miserable comparison the Inspector's
own Report really is, when contrasted with the productions of
men of science and learning.

The Report of Inspector Boole for 1864, dated January
23, 1865, is a comparatively modest document of 53 pages,
of which the Inspector's Report proper fills less than 8 pages,
and is as poor a public document as ever was issued by the
Chief Health Officer of a great city of the world. He no
where gives his Honor the Mayor the *lie*, as in his Report for
1863, but instead of that pungent paragraph of that Report
he says, there " has sprung up an arrogant and self-constituted
association called the "Citizens' Association," and pretty
clearly indicates that that body does not approve, and that it,
with some effect, opposes, those schemes of the City Inspector
that have so alarmed all property holders in the city of New-
York. It is doubtless true that the joint efforts of some
public spirited citizens have somewhat embarrassed the City
Inspector's operations under his "assumed powers."

It is only said of the Mayor, in the Report for 1864, that
not all the urgency of the City Inspector has overcome his
"unwillingness," &c., to "convene the Board of Health," and
hence the "city is deprived of its action *on matters of pressing importance,*" &c.; and hence it appears that the old quarrel
continues, and the health powers, (beyond those belonging
to or "assumed" by the City Inspector,) are, in this time of
public danger, when the city is filled with soldiers going to
the country freighted with disease, *utterly dormant.* The tide
of contagious fever and pestilential small-pox rolls on undis-
turbed from the army to the city, and from the city to all
villages and homes of the State.

The Report for 1864 states the deaths at 25,645 only, being

3

449 more than in 1863; and again there is a like triumphant
comparison of the death rate of New-York with that of Paris
and London; and it is doubtless true that the returns are
just as reliable as those for 1863.

It is a curious indication of the results and of the probable
accuracy of these returns that 1864 shows only 2,637 marriages against 3,272 in 1863, (certainly not indicating a very
favorable tendency of metropolitan morals and wealth,) and
gives 5,592 births against 6,429 in 1863.

The births subtracted from the deaths in 1864, according
to these valuable and accurate Reports, show that the population of New-York in 1864, decreased 20,053 souls. It will,
at this rate of killing off more than 20,000 of the people of the
city annually, above the number that are born in it, hardly
require more time for the City Inspector to depopulate the
place than he is likely to consume in using up all the property
within the city, rapidly as he is known to be advancing in
this latter enterprise. The last man and the last dollar, upon
the joint results of his figures and his action, will soon be
left "alone together."

This Report for 1864 is as silent as that for 1863, about fees
received for the registry of births, deaths and marriages.

A large portion of the document is taken up by a Report
of "L. H. Boole, Superintendent of Sanitary Inspection."
The natural reasons that placed a brother of the Inspector
in the most important position of this great department of the
City Government, which, though this one officer has paid out,
in 1864, (as is admitted by this Report,) the enormous sum of
$812,003.85, has, doubtless, also secured that perfect *harmony
of aim and co-operation* which, it would seem, nowhere else
exists, between the City Inspector's Department and the residue of the City Government. *I repeat, at this moment, the
City Inspector stands suspended by the Mayor.*

The duties of this Superintendent of Sanitary Inspection
are defined in the 37th and 38th sections of the City Ordinances of 1859, as given in the late City Inspector Delevan's

compilation as follows: "He *shall remain, audit and certify* to the City Inspector all accounts for work done under his supervision, and no requisition shall be drawn for any bills, accounts or contracts for cleaning the streets, unless certified by the Superintendent of Sanitary Inspection," * * " who shall, *in all matters, be under the direction, control and supervision of the City Inspector,*" &c. But it should be added that, in beautiful harmony with the last clause, it is also declared (§37) that said Sanitary Superintendent. "shall also *have, exercise and possess all the powers and duties* by law or ordinance conferred on the City Inspector," &c. *That is, one brother may expend* $182,003 85 *per year—the other may approve all the vouchers—the first brother in authority has full control over the last, while at the same time the last has all the powers and duties of the first!* It may be thought, in view of such ordinances, that the well-known economy of those officers has been the only reason why their expenditures have never been thought excessive, and why the people are so well satisfied with the present arrangement!

These Reports do not show much that is pertinent to the question under inquiry. The City Inspector is constantly exercising power under health laws proper, and also under laws relating to sundry other subjects, such as street cleaning, care of markets, &c., &c.; but he does not state how much he expends under the separate heads. He has officers and clerks, who perform both classes of duties, but the time and expense attending the respective classes are not indicated.

At page 63 of Valentine's Manual for 1864, (published at the city expense,) we find a list of 22 "Street Inspectors," who are appointed by the City Inspector. Their pretended duties are, (as defined by the 40th section of the City Ordinances of 1859, still in force, see Morton's Health Laws, p 88,) to report to the "Sanitary Superintendent," (an appointee of the City Inspector,) "the condition of the streets and *all violations of* " *any contract for cleaning them,* and shall each receive for his "services the sum of three dollars per day." Thus nearly

$24,000 is annually paid to this class of semi-Health Officers.
When we know that the streets *are not cleaned by contract at
all*, that these officers are, for the most part, ward politicians,
who are believed to devote their time to their own private
affairs and to politics, we find some further explanation of the
existing public discontent and official corruption and extravagance. The following extract from a Report of a Finance
Committee of the Board of Aldermen, is quite instructive
and pertinent to the matter in hand :

For salaries of the City Inspector and of the Officers,
Clerks, Messengers and Inspectors attached to, or
connected with his office, and in each of the bureaus
and offices in said department,one hundred and thirty-
eight thousand one hundred and sixty dollars......$138,160 00

For salaries of the Resident Physician, Health Commissioner, and the Clerk of the Board, or Commissioners of Health, five thousand three hundred and
forty five dollars.............................. 5,345 00

For advertising, office expenses, and all other expenses
necessarily incurred in the enforcement of the Corporation ordinances relating to said department, not
specified and provided for under other heads, twenty
thousand dollars.............................. 20,000 00

For compensation of the Resident Physician for his
services as agent of the Board of Health, and for
expenses which may be incurred by said Board beyond the amount provided for under other heads of
account, five thousand dollars.................. 5,000 00

Total...$168,505 00

It is believed also, that scarcely any of these Health
Wardens are doctors or are men of much education; that
few of them *neglect their private affairs;* and that all are con-
spicuous as ward politicians, and that they are not wanting
in their admiration of the City Inspector or in promptness in
collecting *their salaries.* But it is said to be generally believed,
that as Health Officers they are for the most part utterly
worthless.*

An examination of the New-York tax levies, as author-

* See the affidavit of Mr. Mulligan, in the Appendix, for startling facts
as to these officers.

ized by the Legislature, for a few years past, will show a gradual reduction in the sums paid to the Board of Health since the subject of a new Metropolitan Board of Health has been agitated. It is yet very suggestive of the corruption of the whole existing system, that the "Board of Health," (that is the Mayor, Aldermen and Councilmen,) have continued to draw their salaries, though it is as notorious as it is disgraceful that this Board has never had a meeting since 1863; nor have the Commissioners of Health, as a body discharged any duties worthy of mention for years. The following are the figures as they appear in the session laws:

		HEALTH COM'RS.
Laws 1859, p. 1125, salary of Board of Health.......$45,000		
" 1860, p. 1017 and 1019, salary of Board of Health. 35,000		$4,250
" 1861, p. 666 and 669, " " " " 35,000		4,250
" 1862, p. 860 and 861, " " " " 6,000		4,250
" 1863, p. 407 and 409, " " " " 6,000		5,345
" 1864, p. 940 and 943, " " " " 5,000		5,345

In 1862, when the salary of the members of the Board were reduced from $35,000 to $6,000, the danger of a change in the law was regarded as considerable; and the further reduction to $5,000, in 1864, indicates that its friends had gained still greater strength.

The salaries of Health Wardens, Assistant Health Wardens, and sundry other pretended Health Officers in the City Inspector's Department are covered by the comprehensive term "*contingencies,*" which abound in all New-York tax levies. Perhaps those officers read with pleasure such entries as these in tax levies:*

1862 p. 861, " salaries City Inspector's Department " $119,228
1863, p. 409, " salaries City Inspector's Department " 119,227
1864, p. 943, " salaries City Inspector's Department " 138,160
1864, p. 941, " contingencies City Inspector's Department "..... 15,000

It appears, therefore, that, during the very years that the greater scrutiny of the public has secured a considerable reduction in the amount avowedly paid to Health-Officers---

---

* Further details as to expenses may be found in the Appendix, under the head of " Abstract of Health Bill."

who do nothing, there has been a very large *increase* secured by indirection to another class of those officers, under the head of "contingencies," and aggregate salaries for the Inspector's Department. Vigilance surely' is not less the price of economy and honesty than of liberty.

Such is a brief view of the laws relative to public health in the great city of New-York, and such are a few of the obvious features of the manner of their administration, so far as they are administered under municipal authority.

There is, however, a branch of the health law applicable to New-York, not yet referred to, which deserves a passing remark. In 1857, the condition of the police law and force of the city were about as deplorable and alarming as that of the health authorities at the present time; and the Legislature, in April of that year, created a Metropolitan Police District and Board of Police Commissioners. The present code of police laws and that admirable force which is the pride, hope and security of the city, have been the happy result. Upon that Board certain health powers have been, from time to time, conferred, partly by reason of its peculiar ability to execute them, and partly by reason of the utter want of confidence in the regular health authorities of the city and their shameful neglect of their duties.

For example, the 29th section of the Police Act of the 10th of April, 1860 (Laws 1860, ch. 259), declares it shall be the duty of the Metropolitan Police Force to "guard the public health."

The exercise of that power is further declared a duty in the 29th section of the Amendment to the Police Act, passed on the 25th of April, 1864; and the Police Board is further directed "to remove nuisances," and to "enforce any ordinance or resolution, &c., applicable to public health." And sections 51, 52, 53, 54 and 55 of the last-named Act go much further, and authorize the Police Board to create a "Sanitary Police Company," and certain limited sanitary inspection is directed to be made by the members of this company. Police surgeons are also provided for, and a limited sphere of duty is assigned these surgeons, mainly relating to the inmates of the buildings and to the men under the charge of the Police Board.

It will be noticed these recent provisions create an additional depository of power in the city over the subject of public health, and some of the powers conferred are of the same nature, and are to be exercised in the same places and in like manner as those still possessed both by the City Inspector and by the Board of Health. Were it not for the inefficiency of the two latter, and that public distrust, which would deprive them of support if they interfered with the Board of Police, the confusion might be increased and serious collision of authority might result.

It is well known that the exercise of health powers by the Board of Police has been far more vigorous, prudent and effectual than by any other authority. And that Board has ample facilities, through its large force, for executing most of the orders of a Board of Health, and with the least confusion and expense. During the past year, the Police Board has cleansed more than 2,500 filthy houses, or has caused their owners to cleanse them. So much has not been done in five years by the City Inspector and Board of Health.

Many subjects that come before a Board of Health demand the consideration of men of scientific and experimental medical skill, as well as the aid of persons of practical executive talent. This fact is understood to be fully appreciated by the members of the Police Board. Executive business talent can best devise effective practical regulations, most promptly remove nuisances, secure the purification of premises and districts, and most efficiently execute orders. But a different kind of experience and knowledge is needed to detect the hidden causes of miasmatic and contagious disease, to guard against the spread of epidemics and the exposures of contagion.

The true composition of a Board of Health in a great city

would therefore seem to be found in a union of administrative ability and executive experience with scientific sanitary and medical knowledge, united with skill and practical experience in the treatment of disease. The Police Board, in a high degree, embodies the former requisites, and members of the medical profession can best supply the latter.

By a union of the two elements, the proposed new Metropolitan Board of Health has been formed.

To prevent a predominance of either element, and to avoid jealousy, the two elements are equally united—four doctors with the four members of the Police Board. The Police Board is made up of equal numbers from the two great political parties; and it is perhaps fortunate that the proposed medical members of the new Health Board are equally divided between the same great parties.

It is certainly desirable to utterly exclude politics from a Board of Health.

It is also highly fortunate that all the eight members of the new Board of Health are willing to serve the public in so noble a cause without compensation. That public spirit of the members of the Police Board, (sustained with some effect, doubtless by a natural wish to get rid of the vexatious annoyances of the heterogeneous health authorities of the city,) which has thus secured their unpaid services, certainly deserves the thanks of the public and the high appreciation of the Legislature. And though the medical profession has rendered of late so much disinterested and uncompensated service in the great cause of public health and morality in the city of New-York, that it is hardly a matter of surprise that greater sacrifices are still offered, yet it is none the less an occasion of congratulation that four distinguished members of that profession, in full maturity of their great abilities and experience, now consent to undertake, gratuitously, so great a labor and responsibility, as to serve on this commission must involve provided only the Legislature will overturn the present disgraceful and inefficient authorities, and con-

fer the proper powers for the protection of life and health
in this metropolitan city. It need only be added, (what is
generally well known even beyond New-York,) that the four
medical men named in the bill are alike in what adorns personal character or distinguishes professional name and position, the foremost in the city and State, and among the first
of the nation and age. They are fortunately able to devote
some time towards removing those great evils of the city with
which they have become so painfully familiar. And frequent visitations of the neglected poor in garrets and in cellars—long and varied practice in hospitals and in families of
every class and condition—familiarity with the highest literature of their profession, and with the writings and doings of
the enlightened Boards of Health which have so much hon-
ored and improved European cities, together with their well-
known practical ability—would seem to qualify these men to
inaugurate the great work of sanitary reform in New-York,
and to afford a guaranty, that what they shall write and what
they shall do, will (in marked contrast with the present administration of our Health Laws and with the Reports heretofore made), secure reasonable safety to public health and be
creditable to the science and learning of the State.

If a Board of Health, thus organized, can be clothed with
the requisite powers, it is believed that their administration
will soon give New-York and the State the blessing of wise
health regulations; and that infancy, now so fearfully deci-
mated, and all classes so needlessly exposed, both in the city
and the county, will soon feel the salutary effects of a great
sanitary reform.

The provisions of the proposed Health Bill are believed
to be well adapted to existing laws, and adequate to secure
the public health.

Its title expresses a leading feature of its provisions.

They are intended not only to preserve public health in
and about the city of New-York, but also "to prevent the
spread of disease therefrom into other parts of the State." It

4

is well-known how readily all infectious and contagious diseases are spread from the city into other parts of the State.

The leading provisions and principles of the bill are the following:

(1.) It creates a Metropolitan Sanitary District, co-extensive with the Metropolitan Police District.

(2.) It makes the four Metropolitan Police Commissioners, (for any time being.) together with the four medical gentlemen named in the bill, a Metropolitan Board of Health ; and all the members are elective for like terms, and are removable in the same manner as the members of the Board of Police, under existing laws. *All the members of the Board of Health serve as such without compensation.*

(3.) As the Board of Quarantine Commissioners has principally to deal with shipping and foreign commerce, and a Board of Health is principally concerned with internal and domestic relations, it was thought best that each should be master of its appropriate sphere, and be independent of the other ; but that each Board should be authorized to co-operate with the other for the promotion of the general public health ; and such are the provisions of the bill, as follows :

(4.) The powers and duties relative to public health, now by law conferred or imposed upon any officer or board in the city of New-York, are, *exclusively*, hereafter to be exercised and performed by the new Board of Health ; and these powers are, in general, the measure of the powers of the new Board, but various additional duties are imposed upon the new Board.

(5.) But as it was understood that representative or influential bodies or officers of Brooklyn, and other parts of the district, were not in favor of having the powers of their respective health authorities materially interfered with or superseded, the alleged wishes of the people of such portions of the Metropolitan District have been respected, and such health authorities can continue the discharge of their functions after the law is passed, as before.

There are strong reasons why all the densely populated region along the Bay of New-York should be under the supervision of one Board of Health ; and (in case of pestilence) it may be more than now the subject of regret that there appeared to be a necessity to sacrifice the real demands of public health, out of respect to the alleged local feeling prevailing in portions of the new district. But the Brooklyn Board of Health is said to be well organized and economically administered.

But certain important general powers, of a kind not interfering with local Boards, are, by the new Board, to be exercised throughout the Metropolitan Sanitary District. These appear in the 12th, 13th, 15th, 16th, 17th, 18th and 22d sections of the bill.

And it is a noticeable feature of the bill that it contains provisions calculated to secure co-operation between all the Health Boards of the State for the dissemination of information concerning the causes of disease, and the means of their removal. The 17th and 18th sections furnish illustrations.

(6.) The Board is to register births, deaths and marriages, without charge, by which the large fees now collected by the City Inspector will be saved, and which, as we have seen, are now never accounted for. Section 13 contains these provisions.

(7.) The subject of street cleaning is incongruous with the duties of a Health Board, and has no proper connection therewith further than this : that when filth is allowed to accumulate, it becomes dangerous to health. A Board of Health should, therefore, have authority to order the proper authorities to clean streets, when that duty is not performed; and this Act gives that authority at section 26, and also the authority to require nuisances to be removed.

Besides, to clean streets would require a large force, the expenditure of large sums of money, and would subject the Board to suspicions of aiming at power and patronage, which might impair its usefulness as an organization set apart to promote the great cause of sanitary science, humanity and benevolence.

There are great abuses known to exist in connection with street cleaning, the full correction of which, doubtless, requires prompt legislative provision ; but all such abuses cannot be corrected by one law, or at one time, and these abuses must be left to their more appropriate remedy.

There are, however, proper provisions in the bill to prevent a duplication of officers and expenses in connection with the execution of Health Laws, and for reporting duties performed under such and under other laws. See section 26.

(8.) The bill provides for the full co-operation of the Police Board and its officers with the Health Board and its officers in promoting public health ; and provides, also, that the Police Board, through its ample means and force, shall execute the orders of the Board of Health. This secures the double result of an economical organization and limited force in the Health Board, and will also prevent collisions between the two Boards in the exercise of their respective powers. These provisions are in the 16th section.

(9.) The 29th section contains stringent and effective provisions for ascertaining, with promptness, any improper diversion of the funds of the Board, and the treasurer is to give ample security. See section 7.

The treasurer may be paid a salary of $500, as provided in the 5th section.

(10.) The organization of the Board is very simple and economical. Besides the members, one of whom is to be president, there may be a sanitary superintendent, with a salary not exceeding $5,000 ; sanitary inspectors, not exceeding ten, with a salary of $1.500 each, and a secretary, with a salary not exceeding $3,500 ; and the services of a sanitary engineer may be employed annually, at a cost not exceeding $5,000. See sections 5, 10, 11 and 22.

These inspectors are to be medical men of several years' experience in the city, and, though few in number, it is thought they will be able to make a pretty thorough sanitary inspection of the city. In London there are forty-eight such inspectors, with numerous assistants.

It is a further important provision that the Board and these officers *shall perform all the duties of the present health authorities, and that, in no event, shall the aggregate compensation of all the officers of the Board exceed two-thirds what is now paid to the existing Health Officers.* See section 11.

(11.) All the other expenses of the Board are *strictly limited* and are placed on the basis of strict economy. No more than $10,000 can be paid for clerk hire; no more than $2,500 for rent; no more than $5,000 for stationery, legal advice and incidentals, unless fitting up offices the first year require that sum to be exceeded. See sections 20 and 27. The only discretionary power of larger expenditure is in cases of pestilence, and in such case no exact limit can be fixed, and in that case, also, the health authorities of Brooklyn and New-York *may voluntarily* co-operate at the joint expense. Section 28.

With the single exception last indicated, the *annual expense of the Board under this bill cannot exceed* $46,000, *and the number of persons in the employ of the Board can hardly exceed twenty.*

(12.) Finally it may be added that there are no experimental provisions in the legal machinery of the bill. There is hardly a section for which there is not a precedent in the Police Bill, as finally matured and amended in 1864.

And it may further be remarked, that the bill provides no opportunities for mischief or extravagance. No money can be squandered; no political patronage wielded; no rights invaded; under it men of a high order of experience and ability, with great disinterestedness are willing, uncompensated, to serve the public in the highest walks of human effort. By this law no valuable organizations are struck down, and only useless and incompetent officers and impracticable legal provisions are superseded.

.I have said it was not a part of my purpose to enter upon the interesting moral considerations presented by this subject, or to narrate the details of disease and death in New-York, as compared with other cities, which have caused so much alarm and awaked so much interest in this bill. But they ought not to be wholly omitted here.

Every Christian will admit it to be a solemn duty of his religion to extend all possible relief to the exposed and the diseased among the humble and the destitute. No intelligent mind will deny that there is a solemn duty resting upon legislators to afford every available legislative aid for the protection and improvement of the health of the people. Disease, filth and crime, are always closely associated. It will justly be everywhere regarded as a disgrace to the State of New-York if her health laws and their administration are not marked by the wisdom, disinterestedness and benevolence which are exhibited, in these respects, in the other great cities and states of the world. And there is too much occasion to fear that this disgrace has been, to a large extent, justly incurred ; for, such brief means of illustration as I have at hand, show that New-York is as far behind the great cities of Europe and America in her precautions against disease as she is before them in the appalling volume and alarming consequences of filth, disease and death, in her midst.

LONDON.—Reference has been made to the organization of a Board of Health in London, in 1848. Under that Board there is a "Sanitary" Police ; and at its head is one of the most distinguished medical men of the United Kingdom ; and forty-eight other medical men are subordinate to him in the employment of the Board. The reports of this Board are valuable documents, and embody the results of the best sanitary and medical skill and experience of England. They explain the best efforts of human effort to control the causes dangerous to human life and health. This Board has control over all the varied nuisances and employments dangerous to health.

When the law creating this Board was proposed, politicians and all those who lived on the spoils of the old system, made the most determined opposition, and, just as has been the case in New-York, they neither wanted funds for bribes, a venal press to defend their opposition, nor a noisy, ignorant rabble to get up a public clamor. A report before me says "they "held public meetings and loudly declared London, for health, "cleanliness, &c., &c., unsurpassed." This is very much the · language of Inspector Boole's Reports to which reference has· been made.

As in New-York, so in London, the friends of reform suffered repeated defeat, but in the end there was a complete and blessed triumph. Of this great triumph of its happy results the London "Times" says:

"It is scarcely credible, but yet. the incontestible figures quoted by Dr. Letheby leave it beyond all doubt, that the average of health throughout the city of London, is higher than the average of health throughout all England, taking town and country together. The mortality in all England is at the rate of 22.8 in every 1,000 of the population : in the city of London, it is at the rate of 22.3 for every 1,000 inhabitants. Gradually the mortality has decreased, until the yearly death-roll of 3,763 has been reduced to 2,904 within the period of nine years, during which the city has been under the rule of the Sanitary Commission. The deaths this year—22.3 per 1,000, or one in every forty-five of the inhabitants—are nine per cent. below the general average, and represent a saving of 286 lives. And, secondly, this gratifying result has been obtained in the face of obstacles which seemed to be almost insurmountable."

LIVERPOOL.—This city affords another illustration of the advantages of an enlightened and efficient Board of Health. When, fifteen years ago, the Board was formed, the death-rate was one in thirty of its inhabitants. A gradual improvement has since taken place and, in 1860, they had, by gradual improvement, reached such a point of healthiness that there was only one death to forty-nine persons. In 1860 *eight* persons died in Liverpool of small-pox, while the same year New-York lost by that disease *thirty* in the week ending

June 29th. Every one is painfully aware of the alarming prevalence of small-pox both in New-York and in other parts of the State of New-York at this time. Liverpool is now one of the healthiest cities of the world.

PARIS.—This well-governed city has an admirable Board of Health. " The Council of Public Hygiene and Health," &c., consists of twenty-nine persons, of whom *fifteen* are physicians, six pharmaceutists, and the residue are engineers, architects and other persons peculiarly qualified. Besides there are local Boards in the city. The *police* enforce the regulations of the Board of Health. The death-rate of Paris is reported to be about the same as that of London and Liverpool, which are far lower than that of New-York. The London and Paris Boards of Health appear to have jurisdiction of a considerable section around those cities.

PHILADELPHIA.—This city has, for several years, had a far more competent Board of Health, with a vastly superior administration, than anything of the kind in the city of New-York. Great sanitary reforms have been effected by such means, and the death-rate has been so reduced, that Philadelphia is said to be the healthiest city of the world of its size.

No man can look over the Health Laws and rules applicable to Philadelphia and the reports of its Board of Health, now before me, and compare them with the Laws of New-York on the same subject and with the pretended Health Reports of New-York officers, without a feeling of shame and humiliation.

The Philadelphia Report for 1863, shows the Board of Health to be composed of eleven persons, of whom four are physicians. There are, in addition, four executive officers, of whom three, if not four, are physicians. The Report contains 55 pages of interesting and instructive statistics, and important and valuable suggestions; all of which show, that the Board is vigilant and industrious in the discharge of its duties. Diseases and deaths are classed, and the causes of prevailing diseases are thoroughly examined; and births, deaths and marriages, are reported. The Report is, in every respect, unlike a New-York Report.

I cannot enter further into details. The document must be compared with a New-York Report to appreciate the difference, and to better comprehend why New-York has a so much higher death-rate than Philadelphia.

By way of illustration, however, a single detailed comparison of this Philadelphia Health Report may be made with the before-mentioned gilt-edged, morocco-bound, boasting New-York Health Report of the same year; bearing in mind that the rates of population in the two cities is about as nine to six:

NEW-YORK REPORT, 1863.	PHILADELPHIA REPORT, 1863.
Reported Deaths...... 25,196	Reported Deaths...... 15,788
" Marriages... 3,272	" Marriages.... 5,475
" Births...... 6,426	" Births....... 15,293

Does any one believe there were, in 1863, 2,203 more marriages and 8,867 more births in Philadelphia than in New-York? or is the explanation of these Reports to be found in the fact, that a reliable and efficient Board of Health has charge of these matters in Philadelphia while neglect, and imbecility, characterize official administration in New-York.

From the fact that so many more marriages and births are reported in Philadelphia it may be concluded, that the reports of deaths are more nearly complete in the Philadelphia than in the New-York; and yet, (omitting small fractions,) *the startling fact is disclosed that in Philadelphia, in 1863, only one person out of every forty-three* DIED, *while, there died in the same year in New-York one person out of every thirty-five.*

This fearful excess of deaths in New-York over Philadelphia is, therefore, either to be attributed to more destructive natural causes arising from soil, climate and exposure in the former city, or to the defective and badly executed Health Laws. To which of these causes do the citizens and Legislature of New-York feel disposed to have this fearful record of disease and death attributed? Can it be that efficient action will be longer delayed at a time when, by reason of a spreading, and fatal contagion, the passage of streets may soon be dangerous, and from fear, pupils, in large numbers, are leaving the public schools?

The Report of the City Inspector of the City New-York for 1861, (not, however, the present incumbent of that office,) furnishes an official answer to the question as follows:

"The causes of this excessive mortality must be sought for in this city, and are readily traceable to the wretched habitations in which parents and children are forced to take up their abode; in the contracted alleys, the underground, murky and pestilential cellars, the tenement house, with its hundreds of occupants, where each cook, eat and sleep in a single room, without light or ventilation, surrounded with filth, an atmosphere foul, fetid and deadly, with none to console with or advise them, or to apply to for relief when disease invades them." He asks, with great pertinency: "How is this state of things, which marks with shame the great city of New-York, to be remedied?" Let those who believe in our present system mark well his reply. He says: "The power of remedy does not rest in me, nor in the department over which I have the honor to preside."

In another report he adds what is fearfully true at this time: "There has been no improvement in the cleaning of the streets, but on the contrary, the city, at this moment, is in a more filthy condition than has heretofore been the case at this season of the year. As an evidence of the effect of this state of things upon the health of the community, I would state that the mortality of the city, has been largely on the increase. Were this increase of mortality the result of an existing pestilence or epidemic among us, the public mind would become justly alarmed as to the future; but although no actual pestilence, as such, exists, it is by no means certain that we are not preparing the way for some fatal scourge by the no-longer-to-be-endured filthy condition of our city."

A late Sanitary Report adds: "It is doubtful if there is a city in the civilized world that contains within its jurisdiction as many sources of that class of diseases known as preventable, or capable of being removed and destroyed, as New-York. They are thickly strewn in every street, in every lane, and even in every dwelling and shop. Many of these sources of public and private pestilence are brought to light in a late Report of Captain Lord, of the Sanitary Company of the Police: ..

"The total number of tenement houses is given at 12,374, with a population of 401.376, of whom 22,095 live in cellars—a subterraneous population large enough for a small city in itself. A little more than two-thirds of the houses, namely, 8,546, with a population of 253.901, are provided with good means of escape in case of fire, while 3,801 houses, inhabited by 125,380 persons, are deficient in this respect. The ventilation of 4,221 houses, containing 141,168 persons, is bad."

Of slaughter houses we have the following exhibit: The number of slaughter-houses in the city is 187, of which 40 are reported as in bad condition. The number of beeves slaughtered per week is given as 2,555; cf small stock, 9,362; of swine, 13,205. Total per week, 25,222. Grand total per year, 1,311,544. Akin to slaughter-houses are those sinks of putridity, fat boiling establishments, of which there are no less than 80 in New-York. These establishments are frequently indicted as nuisances, and are generally located in densely populated districts; but against neither can our health officials be induced to proceed.

PROVIDENCE.—This City has had for some years an efficient Health organization, and has become one of the most healthy cities of the World. Its Chief Health Officer is Dr. Snow, who is a distinguished physician; and some statistics furnished by one of his recent reports may be regarded as impartial and reliable, (even though they do not sustain the theories of City Inspector Boole, nor state anything very flattering to New-York.) He gives the following table of the ratio of deaths to population:

	Estimated Population.	Deaths, 1863.	Of Population, one in.
New-York,	900,000	25,196	35.7
Philadelphia	620,000	14,220	43.6
Boston,	194 000	4,698	41.2
Newark, N. J.,	85.000	1,952	43.5
Providence,	55,000	1,214	45.3
Hartford,	32,000	583	54.8

These appalling facts are further sustained by a Report made by twenty leading physicians of the city of New-York during the past year. They use the following language:

" Previous to establishing a good sanitary government, the annual rate of mortality was:

In London...... ....1 in 20
In Liverpool:.................................1 in 28
In Philadelphia ....1 in 39
In New-York, at present......................1 in 35 +
    Do.   average of last ten years...........1 in 32½

" The rate of mortality in the same cities, with the present system of sanitary government, has been—

In London..1 in 45
In Liverpool...1 in 41
In Philadelphia......................1 in 44 to 1 in 57

" While in the city of New-York the death-rate has increased from 1 in 46½ [in the year 1810] to 1 in 35+ at the present time. ' By means of suitable sanitary regulations, and a faithful and competent administration of such laws, the rate of mortality in this city ought to be very greatly reduced. The experience of other great cities, and the teachings of sanitary science, warrant the opinion that the present rate of mortality may be reduced fully THIRTY PER CENT. Such a reduction would save from 7,000 to 10,000 lives in this city during the present year.

" It is a medical and statistical fact that for every death in a large community there are at least *twenty-eight* cases of sickness. This would give, in the population of our city, *upwards of two hundred thousand cases of preventable and needless sickness every year !"*

With such a fearful record staring us in the face, is it possible that this body should longer delay any practical legislative remedy. The growth, good name and prosperity of a great metropolitan city are in peril; disease and death, the result of neglect, filth and inefficiency, threaten us on every hand. In a city of near a million of people—the great centre of wealth, refinement, elegance, medical skill, benevolence and Christian effort of a great nation—garrets and cellars are filled with contagion and death--lanes and alleys are poisoned with nauseous and dangerous emanations—sewers and open places are filled with germs of endemic disease— the odors of decaying animals and miasmatic filth spread along the avenues where liveried servants drive the coaches of the rich, and enter the churches where the humble and the proud kneel at the altar. No parent or child can now visit

the city from the interior of the State without awakening
anxious solicitude for a safe return. Every one who enters a
stage or car is anxious lest his neighbor be a source of contagion. *Years have elapsed since there has been any meeting of a
Board of Health;* and measures calculated to protect life and
health are abandoned at a period of the greatest danger; the
Mayor refuses to call any meeting of the health authorities,
because he deems them dishonest or incompetent; the City
Inspector, though under sentence of suspension from office
by the Mayor, spends near a million a year, and yet leaves the
streets of the city in a condition that would disgrace the camp
of a tribe of Kalmuc Tartars. Yet such is the general leth-
argy or despondency among the mass of the people in regard to
all measures of reform, that the very officer who, in his official
Report for 1863, boldly proclaims that he "has assumed responsibilities unsanctioned by municipal ordinances"—an
officer who, day after day, has been standing at the bar of a
a committee sent down by the Senate in answer to the appeal of the city against his unblushing abuses and stupen-
dous extravagance—that officer dares, in his Report for 1864,
to attempt, by a low and vulgar appeal to the honest but ignorant prejudices of the large class he hopes by his use of
public money to control—to attempt to intimidate those generous and public-spirited men who are endeavoring by disinterested co-operative effort, under the name of the "Citizens'
Association," to achieve something in the great cause of
municipal reform.

It is no part of my purpose to attempt any defence of that
Association, or of its numerous members. Some results of
their labors will, I believe, be before the Legislature during
this session, which will speak in their behalf. Certainly no
charge of the City Inspector will affect their high standing
with the better classes in New-York.

The bill before us rests on no association, except that happy
association which a wise Providence has ordered shall always
exist between disinterested Christian effort in behalf of suffer-

ing and imperilled humanity, and its natural fruits, which are prosperity, health and happiness. The need for this law is older than any "Citizens' Association," and though it may be much indebted to the support of that body, the same need will not be diminished, even if that "Association," should unfortunately falter in its great undertaking, without accomplishing what it has so worthily begun.

It will still remain the duty of the philanthropist to labor in behalf of imperilled life and health—it will still be the interest of the merchant and the banker to aid in removing that disgrace and danger which makes the traveller approach New-York with dread—it will still be needful to have legislative enactment to secure in New-York that sanitary reform which the meanest instincts of fear, and the lowest forms of state pride, as well as the solemn invocations of the pulpit, and humble prayers of mothers in cellars and garrets, at the bed-side of children sinking needlessly to untimely graves—alike demand.

But it may be said that mere legislation changes cannot produce the needed reform. That what is demanded is better officers, more disinterested public spirit in the city, and that exercise and guardianship of the elective franchise which would secure a fuller execution of the laws. It cannot be denied that if such virtues and vigilance more generally prevailed, great reforms would follow. But it sometimes happens that laws are made on the suggestions of bad men, and that they are, without the cognizance of the Legislature, framed so as to facilitate dishonest purposes. I think it could be proved that such has been the fact in regard to some of the enactments applicable to the city of New-York.

Until after 1857, the better class of men in the city were not much aroused to the necessity of watching and combining to counteract the deep-laid schemes of knaves and professed politicians. Since that date the conservative and honest influences have, to some extent, combined against those two mischievous classes, and have succeeded in achieving a great triumph in the establishment of a well-disciplined Metropolitan Police. They have also thoroughly aroused the public mind to the necessity of reform in other quarters. The contest for a Metropolitan Health Law has been waged during the same period, and though its friends have sustained repeated defeats, they are, this year, much stronger than ever before.

They are not, however, unmindful that money which (it is believed) defeated them in 1861, and which, aided by all the arts of intrigue, prevented a final vote in the Senate, in 1862, may be used against them in 1863. There are the same interested combinations—the same officers who draw salaries without performing duties—the same specious arguments, founded on alleged invasions of municipal rights—the same pensioned scribblers who silence the disinterested utterance of portions of the public press—the same owners of city nuisances, who fear their noxious occupations might not be tolerated—to contend with now, who have, every former year, combined against the advocates of an enlightened Health Law.

But the friends of the bill are confident that, if their arguments are not fairly answered, no unworthy influence can, at this time, prevail.

The friends of this bill do not invade any municipal rights. They only aim to sustain the honor and to promote the best interests of New-York—to lift into places of power and high responsibility the worthiest citizens of that great metropolis— and regard it both appropriate and necessary to invoke the aid of the Legislature in such a cause. Disease, contagion, pestilence, are not bounded by city or county lines—cannot be circumscribed within constitutional limits or political divisions. They travel along our railroads and our rivers— they fly rapidly with the winds, over the hills and along the valleys—they are borne with the parcels from the merchant, and even in the sealed missives of friendship, to every family fireside in every portion of the State. And to assert, for the city of New-York, the right to allow its streets and hotels, daily and nightly thronged with the precious lives of every town and village, to become poisoned with fatal exhalations and pregnant with disease and death, is to insult the general intelligence of the State, and to mention that life and health are something less sacred than a political dogma and ought to yield to the claims of a pretended municipal franchise.

Is it longer to be tolerated, that while other great cities send forth instructive and learned reports, written by able physicians, full of valuable information and disclosing wise measures for the protection of life and health—reports credit-able to science and humanity, and everywhere read and pon-dered—the health authority of New-York shall remain un-exercised, or be exerted and illustrated only in acts and re-ports that are a stain upon its honor, a scandal to its science and a disgrace to its literature.

What is demanded is, that there should be a commission composed of educated, independent and practical men, deriv-ing their power from the State and independent of popular suffrage in the city—men who will have the courage, power and judgment requisite to deal with all those causes that now peril life and health.

All the Health Officers of the city now derive their author-ity either directly from popular suffrage or from the Mayor and Board of Aldermen. The Mayor is elected for two years, and is sure to be a candidate for re-election. The City Inspector and Resident Physician are nominated by the Mayor and confirmed by the Aldermen. There are seventeen Al-dermanic Districts in the city, and the Aldermen are elected for a term of two years.

There are six Councilmen, elected annually, from each Senatorial District, for a term of one year.

The men who keep groggeries, tippling shops, fat-boiling establishments and butcheries—who own dance-halls and houses of ill-fame—who allow cellars and places to be filled with stagnant waters—who neglect drainage and purifications

--who build ill-ventilated tenement houses—who keep filthy and offensive stables—all those who, in a thousand other ways, contribute to make New-York the most unhealthy of the large cities inhabited by a civilized and Christian people—are themselves, with all their friends and dependants, *voters*—the constituents of these Aldermen and Councilmen—are, in all probability, the most active politicians of their respective wards, whose support or opposition makes or unmakes some city father. The Police Justices are, in a similar manner, dependent on the same class of men; and convictions before them, for nuisances, are known to be of rare occurrence. Is it supposed that the officers whom such men have elected will, when in office, interfere with the doings of their most active constituents—proceed against the very premises as being a nuisance, or dangerous to health or morality, in which, perhaps, the political caucus was held that secured their nomination?

No one who knows anything of New-York municipal administration, will expect anything of the kind. These officers, if public suspicion be correct, are, some of them at least, much more likely to be the champions and the apologists of the offences and offenders against public health in their respective districts; and those who should be punished, are quite likely to be on the long sinecure pay-rolls of the City Inspector, side by side with the hireling scribblers for a portion of the city press, who receive so much public money for palliating official delinquencies, that they are enabled to serve their literary employers at rates most agreeably moderate. It may be that that class of writers may sometimes be kind enough even to report the doings of this Honorable Legislature on the like moderate terms, and, forsooth, include therewith elaborate and pungent abuse of all friends of reform, without additional charge. It is certainly convenient to receive a salary from the City Inspector, and to draw one's pay as a "reliable" correspondent, but it might not be pleasant to have the facts made public, or creditable to any party concerned. For such abuses and sources of corruption, there is no adequate remedy but a law similar to the one now proposed.

And, therefore, against every obstacle, and after every defeat, the friends of this great reform, relying on the justice of their course, sustained by a high sense of duty, which is its own reward, are firmly resolved to persevere until final triumph shall crown their efforts.

In behalf of the neglected dying poor—by all the priceless value of life and health, now doubly imperilled by diseased soldiers from the army, and by a fatal contagion now disclosing itself in so many of our cities and villages—in honorable emulation of the great work of sanitary reform so much advanced in other American and European cities—in response to the generous offer of service by those who are ready to execute this high commission without pecuniary reward—in harmony with that regenerating public sentiment which is purifying the national health and life—in obedience to all the holy promptings of virtue and religion—by one effective, worthy effort to rebuke official imbecility and corruption, and inaugurate benevolent, practical reform in the city of New-York— let this Legislature honor itself and bless the whole people of the State by enacting this great sanitary and all-pervading measure of reform. Future generations will wonder at the opposition it encountered, while they rejoice in the priceless benefaction.

APPENDIX.

§ 1. Makes the Police District a Health District.

§ 2. Makes the four Police Commissioners and four Doctors, the sole Board of Health in New-York.

§ 3. Names Dr. Alonzo Clark, Dr. Willard Parker, Dr. James R. Wood and Dr. Isaac E. Taylor, as the four Doctors and Health Commissioners, with the Police Commissioners.

§ 4. Makes the terms of office, after first Board, eight years, and provides for vacancy.

§ 5. Provides for election of President, Treasurer, and Secretary.

§ 6. Defines duties of President and Secretary.

§ 7. Relates to Treasurer, and his bond.

§ 8. Forbids members of Board holding political office.

§ 9. Authorizes Governor to remove members of Board.

§ 10. Authorizes Board to create "Sanitary Superintendent," defines his duties and fixes his salary.

§ 11. Authorizes Board to appoint "Sanitary Inspectors," not exceeding ten, fixes their salary and defines their duties.

§ 12. Defines the powers of the Board, giving all (but limiting the same to) the powers now distributed to and through various offices and departments, viz.:

1. The Mayor.
2. The Board of Health.
3. The Commissioners of Health.
4. The Resident Physician.
5. The Health Commissioner.
6. The Medical Agent of the Board of Health.
7. The City Inspector.
8. The "Sanitary Superintendent," the 22 Health Wardens and 22 Assistant Health Wardens, and the other numerous subordinates of the City Inspector's (Boole) Department.

§ 13. Gives the Board charge of the Registry of Births, Deaths and Marriages.

§ 14. Prevents conflict, and provides for proper co-operation between the new Board, and the Quarantine Board and the State Health Officer.

§ 15. Provides for action of Board out of the City, and co-operation with Health organizations in the District out of the City in time of great peril.

§ 16. Requires co-operation with Police Board and officers, to protect Health, and for the execution of the orders of the Board of Health by the police force.

§ 17. Directs the Board to preserve valuable facts, and give useful information to Boards of Health in all parts of the State.

§ 18. Provides for an annual report from the Board to the Legislature, and for sending a copy to each Board of Health in the State.

§ 19. Authorizes by-laws and directs copies laid before the Legislature.

§ 20. Authorizes clerks, but the whole cost not to exceed $10,000 ; and allows incidental expenses for stationery, light, fuel, &c., not over $5,000.

§ 21. Authorizes keeping of public " Complaint Book."

§ 22. May employ Sanitary Engineering service, at not exceeding an annual expense of $5,000. Health officers may inspect works, grounds and buildings.

§ 23. Provides for meetings of Board.

§ 24. Directs Board to aid in enforcing Health inspection laws ; to demand reports from dispensaries, hospitals, &c., &c.

§ 25. Provides that the Board shall not be required to make sundry reports, now only required by reason of the powers over Health being distributed between so many officers, departments and agents.

§ 26. A long section, made necessary by reason of the present confusion in laws, and of the fact that the same officers who have charge of public Health proper, have also the control of markets, weights and measures, street cleaning, &c., &c.

These matters have no necessary connection with Health, unless they are neglected, and when they are so neglected and thus become nuisances, power to order the proper authorities to perform their duties is given to the Board of Health by the Bill.

The last clause of this section requires the City Inspector to file a report as a check on his expenditures, and for future information.

§ 27. Authorizes rent of office at expense not exceeding $2,500.

§ 28. Authorizes joint action between Board of Health in Brooklyn and New-York, in case of pestilence, and provides for expense.

§ 29. Provides the money to pay expense of Board shall be paid into State Treasury annually ; also provides for *summary examination of any member of Board where any division of funds is charged.*

§ 30. Provides for raising money for paying expense of Board, being the same provision that is contained in the police law.

§ 31. Authorizes the borrowing or raising of funds for first year in anticipation of taxes.

§ 32. Makes violations of Act a misdemeanor.

§ 33. Authorizes its records to be used on judicial proceedings.

§ 34. Makes Act take effect June 1, 1865.

A FEW SUGGESTIONS MAY BE USEFUL.

1. It will be observed that more than *one-third* of the length of bill is occupied with drawing the boundary between such powers of certain officers and departments, as are and as are not given to the Board of Health. It was certainly not proper to give the general *legislative* powers of the Common Council, now with the Mayor, a Board of Health, nor was it consistent with the duty of a Board of Health to have charge of 1,000 men, or $500,000 to clean streets. It requires long and carefully drawn articles to define what portion of such and various other powers now conferred on the same officers in the greatest confusion shall be given to a properly constituted Board of Health.

2. It was not desirable to interfere with the efficient Board of Health in Brooklyn, nor with the Quarantine Board; and yet, it was important that the Metropolitan Board of Health should have certain important powers *throughout the district.* (not before conferred on any Board, which are given in § 14, 15, 17, 18, 22, 24, 23); and to preserve the two Boards referred to, and give the general powers needed, has required large space in the Bill.

3. Had not the foregoing facts presented obstacles, the Bill would not be half its present length; and those facts do not affect the organization of the Board, which is very simple and well adapted for prompt, harmonious and efficient action.

(1.) The eight Commissioners act as a single Board, of which (§ 9) five are a quorum, and there is no occasion to consult any other Board or officer.

(2.) The Secretary will keep the records, and conduct the correspondence of the Board.

(3.) The " Sanitary Superintendent " is the chief executive officer of the Board, and has a general superintendence, as the Board by its by-laws shall direct.

(4.) The ten " Sanitary Inspectors :"—They will daily examine into causes endangering Health, under the charge of the superintendent.

(5.) Any engineer the Board may for any occasion employ, in connec-

tion with sewerage or otherwise, will attend the work assigned
him, as the Board may direct.

(6.) The police force *will execute all orders of the Board*, except such
as the inspectors or superintendent may personally execute.

4. It will be seen, that the expense is reduced to the lowest possible
point, and may be shown as follows :

(1.) The members of the Board serve without pay.
(2.) Secretary, not to exceed..................$3,500
(3.) The Sanitary Superintendent.............. 5,000
(4.) Services of Sanitary Engineer.............. 5,000
(5.) Ten Inspectors, $1,500 each...............15,000
(6.) Clerk hire, not exceeding..................10,000
(7.) Rent of offices, not exceeding............. 2,500
(8.) Incidentals, Stationery, &c., &c............ 5,000
(9.) Pay of Police Force (not increased).........

Total...............................$46,000

5. The Bill everywhere follows the analogy of the police law, and
contains no experimental legislation.

6. The great difficulty in New-York was, not to any great extent,
the want of power over the subject of Health, but grew out of two leading facts :

(1) The Health officers are inefficient, and have not the proper qualifications, *and the mode of their election is such that better ones
cannot be secured without a change in the laws.*

(2) The existing laws contain conflicting *provisions and the power
over health is deposited in so many offices and Boards that co-operation or harmony in their execution has been found unattainable.*

Hence a *concentration* of power in proper hands rather than additional
power is the remedy sought. But valuable *additional powers* over the subject of small-pox and other pestilential diseases have been added to the
12th section of the bill since it was first printed. They were taken from
an efficient law long in force in Brooklyn.

7. Contrast this organization with the existing organization and expenditure.

The present organization is as follows :

(1) Any ten of { 17 Aldermen and 24 Councilmen, and the Mayor. } make a "Board of Health."

(2) The President of the Aldermen,
The President of the Councilmen.
The Health Officer,
The Resident Physician,
The City Inspector,

constitute a second Board called "The Commissioners of Health."

(a)*

(3) The City Inspector—This officer claims and exercises at this time nearly all the authority over Health, (now practically known in the City,) under Tit. 3, Chap. 275 of the Laws of 1850.

(4) The Health Officer.

(5) The Resident Physician, } appointed by the Mayor, with the
(6) The Health Commissioner, } consent of the Aldermen.

(7) The Agent of the Board of Health, appointed by the Board of Health. The *same person* is said to be at this time both Resident Physician and Agent of the Board of Health, and to draw salary in both capacities.

(8) The Health Wardens, *twenty-two* in number. } They are appointed
(9) The Assistant Health Wardens, *twenty-two* in number. (a) } by the City Inspector with the consent of the Aldermen.

(10) The "Superintendent of Sanitary Inspection," a subordinate of the City Inspector.

(11) "Registrar of Records and Statistics," another subordinate of the City Inspector. (b)

(12) The Mayor.

(13) Numerous Clerks of the sundry Boards and officers above mentioned. (c)

8. Expenses of present system. It is not easy, owing to the reticence of officials and the method of making up reports, to get at the salaries of the foregoing Health Officers. As to some there is no doubt, and as to the residue a pretty close approximation and something like certainty may be readily reached.

(a)* The Corporation Council has decided, and it is generally believed, so far as there is any harmony, of opinion on the subject, that each of the two Boards has the same authority, and as there are 41 Councilmen and Aldermen, of which *any ten*, with the Mayor, make a Board of Health, it will be seen *that those two bodies furnish material for four Boards of Health in addition to that made up of the Commissioners of Health.*

(a) See affidavit attached as to the qualification of those officers and the influences that secure their appointment.

(b) This officer gets up the bulky reports from the City Inspector's Department, and is put forward as one of the chief Health Officers of the City ; though in fact he is a mere *clerical officer* and has no authority worthy of mention.

(c) It may be remarked that it is a great question with the citizens of New-York whether this "Registrar," the "City Inspector," the "Resident Physician," the "Health Commissioner," or the "Agent of the Board of Health," is the *greater* of the aforesaid officers, but *if one doctor holds two of these offices*, as is reported, he would seem to be clearly entitled to precedence.

(1) Board of Health proper, (that is sums paid Aldermen,
Councilmen and Mayor as a Board of Health.) (a)......$35,000
(2) Board of Commissioners of Health; their salary as a
Board, the writer has not satisfactorily ascertained, but
estimates it at.................................. 12,000
(3) The City Inspector............................ 5,000
(4) Health Officer, unknown.
(5) The Resident Physician........................... 1,250
Same person in capacity of Agent of Board of Health,
estimated $4,000.............................. 5,250
(6) Health Commissioner........................ 3,500
(7) Agent of Board of Health, as above.
(8) Twenty-two Wardens from $1,000 to $1,200 each...... 25,000
(9) Twenty-two Assistant Wardens, $1,000 to $1,200 each. 25,000
(10) Superintendent Sanitary Inspection, (estimated)....... 5,000
(11) Registrar of Records, &c., (estimated).............. 4,000
(12) The Mayor, (unknown.)
(13) Clerks, (estimated)............................. 10,000
(14) Rent, incidentals, sundries, &c., &c., (estimated not less
than) 15,000
                                            ————
                                            $151,750

It will, therefore, be seen, even if the entire salary of every officer who performs any other duties than those under Health laws proper, be rejected, and if no salary at all be allowed the Commissioners of Health as a body, and if the Salary of the Board of Health be only the sum appearing in the Statute of 1864 ; and if the incidental expenses estimated for, (which are believed to be far too small,) *are reduced by half; that after all these deductions are made, the expenses of the present system will yet be shown to be nearly double the amount required under the new law proposed.*

For proof that the present expensive organization is utterly inefficient and is incapable of reformation, see the extracts from the report of Mr. Delavan, late City Inspector, printed in this Appendex.

NOTE B.—DELEVAN'S REPORT FOR 1861.

The following extracts from the report of the late City Inspector, Delevan, the immediate official predecessor of the present Inspector, (F. I. A. Boole,) afford a most significant answer to most of the representations of the latter officer and of his subordinates, and confirm the allegations of

(a) See Laws 1960, p. 1017 and 1019, and Laws 1861 p. 608 and 659, where that sum is so appropriated, though in 1864 it was apparently reduced to $5,000, Laws 1864 p. 940 and 943, by reason the efforts to obtain a new Health Law.

the friends of the pending bill, as to the distressing need of reform and the hopelessness of any relief under the present system. The report quoted is Mr. Delevan's official report, as City Inspector, (a pamphlet said not to be easily obtained at the present City Inspector's Office just now,) dated January 13th, 1862. Having read these remarks of Mr. Delevan, (for the frankness of which he perhaps lost his office,) a more correct estimate can be formed of the consideration due to the allegations of Mr. Boole now before the Legislature; which seem to imply that his department is a model of efficiency—that the Health organization of the City is wonderfully efficient—that the laws are beautifully harmonious—that the health of the City of New-York is admirably protected—that the Board of Aldermen and Councilmen are the most disinterested, the best instructed and the most independent bodies that could possibly be selected to appoint Health officers and devise and enforce Health regulations. Is it not strange, indeed, that a man with such plans and opinions as Mr. Delevan should have been compelled to give place to Mr. Inspector Boole and his theories?

Speaking of the defects of the registering of births, deaths and marriages, Mr. Delevan says, " Nor can this defect be remedied unless by legislative interposition, and by the passage of new laws with suitable penalties for non-compliance, which shall compel an obedience, heretofore reluctantly yielded, and frequently refused." Page 9.

" But this accuracy, easily attained as regards the death record, cannot possibly be reached as regards the record of the births and marriages, in consequence of the defects in the present statute to which I have adverted. Thus one of the principal objects for which the Bureau of Records and Statistics was instituted is defeated, and the usefulness of that branch of the department seriously impaired." Page 10.

Speaking of the unhealthy and neglected condition of the City, he says, " I refer to the under-ground apartments, the hot beds of febrile disease, where it exists in its normal condition, ready for all subjects that are placed within its reach. Nothing can be more idle than to expect that those who dwell in these vaults of sickness can remain in the enjoyment of health. The Medical Faculty, with one accord, join in this opinion, and the question is forced upon us whether it shall be permitted to landlords to let such tenements to lodgers." Pages 23 and 24.

At page 25, he says, " In the upper part of the City, where the population is rapidly increasing, the fatal effects arising from the miasma of sunken and unfilled lots from the pools of stagnant water with which many of them are filled, are to be seen in the records of our annual mortality."

Again, page 25: " I refer to the system of tenement houses, which has sprung up so gradually among us, that its increase has hardly been noticed.

7

An estimate has been made which places us in possession of the information, that the mortality of those occupying these abodes, including infants
and children, is as one to twenty. There seems to be no doubt of the
correctness of this estimate, or, if there be any error, it is that the propor-
tion of deaths is larger than herein described."

Again, page 26 : "It is frequently the case that these buildings cover
whole blocks, and such are the imperfections for the admission of light and
ventilation, that, in some places, the sun is almost entirely excluded, and
the air finds admission only from the doorway or the narrow passages leading to the rooms."

Speaking of the necessity of new legislation, he says, pp. 56 and 57 :
"It is interesting and instructive to trace the movements of the City
authorities after their first earnest attempts to protect the Health of the
City, and their efforts in a new vocation; in which the predominant trait
seemed to be an *utter want of knowledge of all the principles and laws which
form the basis of sanitary success.* But if their movements were uncertain
and their success limited, there is nothing strange in such a condition of
things, *for a moment's consideration must convince us that qualifications
necessary to conservators of the Public Health, are rarely, if ever dreamed of in
the selection of members of our City Councils.* However just may be their in-
tention, or with whatever zeal they may be disposed to dedicate themselves
to this incidental but most important branch of public duty, *what portion of
the community ever dreams in their selection of their qualifications for the most
responsible trust of guarding the Public Health ?* So little can be expected of
them in the discharge of this branch of their duties, that it would be almost
unjust to condemn them for a dereliction of those duties, delicate and vital as
they are, *the performance of which never entered into the minds of those who
elected them. It may be safely said, that the selection of candidates for our
City Councils, on the ground of their fitness for the preservation of the Health
of the City, is a thing almost unknown.* Other tests, whether on partisan
grounds, or on higher grounds of the welfare of the City may be required,
but this test surely not the least important, has been and will remain
unheeded. If then, the operation of the *total inability of our local legislators to handle a subject, the details of which have never engaged an hour of
their lives, be admitted,* the charge of dereliction of duty on their part in
the hour of danger should not be harshly made, and there should be no
disappointment of expectations on that score, for no such expectations
were ever entertained by those who placed them in power. *From the very
nature of our elections there can be no improvement expected in the present
state of things as regards the Public Health, so far as the Common Council
are concerned ; and the remedy apparent to every one must consist in the*

adoption of laws transferring the power of sanitary regulations to some other authority, of a different order of instruction in sanitary science."

And, further, at page 58 to 60, he says:

"The Common Council, first acting in one capacity, as legislators, under the disadvantages to which I have called your attention, those advantages become doubly apparent when they are resolved into an exclusive body, to act on one subject alone, and the *whole process becomes characterized with an inefficiency which would be regarded as ludicrous, were it not that the subject is of too vital importance to be lightly dealt with.* When the matter has thus been bandied about, *it comes up for the third time in the body where it originated—the circumlocution having run its round—and ready for a new start, to travel in the same circle where it began, having, in the mean time, not made the slightest advance towards the preservation of the Health of the City—* the warding off impending, or allaying actual pestilence. *Such is the construction of the Board of Health, one branch of our sanitary system;* and the other, the Commissioners of Health, *is open to equal objection, from the manner of its organization and the want of a sufficient power.* This body is composed of six persons, besides the Mayor, who is a member, *ex officio.* It consists of the President of the Board of Aldermen, the President of the Board of Councilmen, the Health Officer, the Resident Physician, the Health Commissioner, the City Inspector. It will be seen that the Common Council is here again represented by two of its members, and that the Board has the benefit of three medical officials, *but, not having conferred upon it the full powers possessed by the Board of Health, it cannot, in many instances, carry into effect the measures deemed by it necessary for the preservation of the public health.*

"The Board of Health, *from the 1st of January last, down to the present day, has not held a single meeting to consider the sanitary necessities of the city. With such a system, can there be a wonder that the sanitary condition of the city is not improved, or that the City Inspector is crippled in his power of action in this respect?* Nor must the consideration be kept from view, that the members of the Common Council, the Board of Health, and Commissioners of Health are all, *from the manner of their appointment, subject to partisan influence. To expect a perfect sanitary system under such a condition of things is to expect an impossibility.* The first groundwork of reform, in the opinion of the undersigned, *is to bestow upon some other body, differently constituted, all power over the sanitary affairs of the city; and until this is done, all other proposals of reform will be deprived of their essentially beneficial features. To escape present complications is the first great point to be gained; and, this point secured, simplicity, promptness and efficiency may be substituted for inefficiency, complication and delay."*

Again, at page 63, he says :

" As an additional measure of reform, I would recommend the abolishment of the present system of Health Wardens, and the consolidation of these offices in the Visiting Physicians of the Dispensaries, who are to receive, as compensation, the pay now allowed to those Health Wardens. In effecting this change, the city will have the safest security that its interests will be attended to, with promptness and efficiency, by a body of men in all respects qualified for their duties."

Again, at page 67, he says :

" It can hardly be questioned, that even the members of the Common Council would gladly be relieved of a burden that does not legitimately belong to their legislative functions—a responsibility from which, while they cannot shrink, they yet feel incompetent to assume ; which brings with it no rewards for duties well performed, and is almost sure to bring condemnation when those duties have been imperfectly discharged, notwithstanding they may have been so discharged with the best motives."

NOTE C.—MULLIGAN'S AFFIDAVIT.

HEALTH WARDENS.—1864.

	Names.	Occupations.
1st Ward	Michael St. George	Liquor Dealer.
2d "	Charles A. Lamont	Refiner.
3d "	John J. Taylor	Warden.
4th "	Daniel Leamy	Notary.
5th "	James Lawrence	Warden.
6th "	John Donnelly	"
7th "	James Kennedy	" Ship Caulker.
8th "	George Cox	Oyster Saloon.
9th "	Alva Terhune	Warden.
10th "	Roderick T. Entwistle	Liquor Dealer.
11th "	Alois Ludwig	Warden.
12th "	James Hope	Liquor Dealer.
13th "	Samuel O. Donnell	Warden.
14th "	John Cavanagh	Liquor Dealer.
15th "	Elwood T. Jones	Mason.
16th "	Peter Welsh	Carpenter.
17th "	John Forney	Warden.
18th "	John H. Forman	Surveyor.
19th "	Patrick Carroll	Liquor Dealer.
20th "	Joseph Brennan	Warden.
21st "	Patrick Dee	"
22d "	Thomas Higgins	Liquor Dealer.

53 ·

ASSISTANT HEALTH WARDENS.

Names.		Occupations.
1st Ward	Andrew Carey	Warden.
2d "	Walter Joyce	"
3d "	John J. Crotty	"
4th "	John S. Roche	"
5th "	James O. Hall	"
6th "	Terence Foley	Liquor Dealer.
7th "	James Leo	Liquor Dealer.
8th "	Ralph Bogert	Trunks.
9th "	Thomas Culkin	Warden.
10th "	Asa H. Bogart	"
11th "	Henry Wolthman	"
12th "	Isaac Vermilyea	"
13th "	James Davison	"
14th "	Patrick Barnes	Clerk.
15th "	Epes E. Ellery	Paints.
16th "	John Ready	Liquor Dealer.
17th "	Patrick Brady	Warden.
18th "	Ananias Matthews	"
19th "	Philip Fitzpatrick	Liquor Dealer & Builder.
20th "	Herman Eivisch	Warden.
21st "	William Reynolds	"
22d "	Alexander Wilder	Notary.

The following persons have been appointed for this year, 1865 :

Occupations as found in Directory.

21st Ward	Jeremiah Crowley,	Warden,	Mason, 170 East 34th St.
8th "	Ralph Bogert,	"	
15th "	William Stevenson,	"	Soap, 186 Laurence St.
21st "	John Murray,	Ass't Warden,	
20th "	James Casey,	Warden,	Deputy, 335 10th Av.
3d "	James McClosky,	"	Boots, 257 Geeenwich St.
20th "	Adam Moser,	Ass't Warden.	}
6th "	Michael McManus,	Warden,	} Cor. Worth & West B'way, Liquors. } Not Found
16th "	Charles Warwick,	Ass't Warden.	}

COUNTY OF NEW-YORK, ss.

James Mulligan being duly sworn, deposes and says : That he resides in 117th street, between 1st and 2d Avenues, in the 12th Ward of the City of New-York, and has for over sixteen years resided in said Ward ; that the foregoing list of Wardens and Assistant Wardens of New-York, for the past and present year, and of their occupations, he believes to be correct.

That many changes take place during the year, and he cannot state the length of time said persons held the office. That he has made personal examination and inquiry as to nearly all said persons, and has ascertained said persons, so inquired of, now hold said office, and believes all of them so indicated now hold said office, except as hereinafter mentioned.

That those mentioned as " liquor dealers," are keepers of stores or cellars where liquor is sold at retail, and as to some of them he knows, and as to all he believes, they personally and regularly attend to their daily and nightly customers at their stores or cellars. These places are known in New-York as " corner liquor stores," and are the resort of large numbers of intemperate persons.

And deponent further says, that it is commonly reported and understood in New-York, that liquor dealers obtain and retain the position of Health Wardens or Assistants by reason of their influence on such of their customers as are voters.

That James D. Hall, Assistant Health Warden of the 5th Ward last year, is put down in Valentine's Manual as residing at 179 Church street, and on going there, deponent finds 179 Church street to be the house of 27 Hose Company, and said Hall reported to be a " bunker " and exempt fireman of that Hose Company, (that is, he resides and sleeps in the engine house, though not on the list of active firemen.) His name is not given in the City Directory.

Patrick Brady, named in said list, was Assistant Health Warden last year, (and is at the present time,) in the 17th Ward, and reported by the Chief Engineer to be a member of Hose Company No. 17.

That deponent is informed and believes that Philip Fitzpatrick, named in said list, as Assistant Warden in the 19th Ward, during 1864, built a row of houses, running from 48th to 49th street, on Sixth Avenue, and, as such builder, he could control a large number of votes. That it was currently reported at the time of his being made an Assistant Health Warden, that he was made such to secure for the Aldermen of the Ward the votes he could influence, and deponent has reason to believe, and does believe, such to be the fact. In the City Directory for 1865, said Fitzpatrick is designated as builder and *liquor dealer*. That one James Smith is reported as now nominally performing the duties of Assistant Health Warden for said Fitzpatrick.

The Health Warden and the Assistant Health Warden of the 19th Ward at this time, (unless they have been changed within a few days,) are the keepers of the most extensive, and most frequented, and notorious liquor shops and groggeries of said Ward ; and these facts are perfectly notorious in the Ward. Liquor is sold at these places, at retail, to men, women and children. That the liquor store of Carrol, the Warden of said 19th Ward, is at the corner of Third Avenue and 42d street, and is a second rate grog shop, and the open resort of the idle and intemperate of the Ward, and said Carrol is an active politician.

That deponent knows the Warden and Assistant Warden of the 12th

Ward of said city very well. The first, James Hope, keeps and personally attends a liquor store, and the last (Isaac Vermilyea) is not, as deponent has reason to believe, in any business but politics, and is to be found generally at Hope's liquor store; and during the time said Assistant Vermilyea, has held said office, he has devoted but very little, if any, time to the duties of his office, and causes of danger to health in said Ward are almost wholly neglected.

That the liquor store of said Warden, (Hope,) is the political headquarters of the 12th Ward, and (for the years 1863 and 1864), the Alderman of the 12th Ward, (Jacob M. Long,) made Hope's liquor store or bar-room the political head-quarters of the 12th Ward politicians, which he represented. That the primary elections and meetings in the Ward were always held in Hope's parlor, over the bar-room, or in the back room adjoining the same.

And deponent further states, that he has acted as secretary of such meetings, and is acquainted with the manner and place of conducting them in other Wards, and he believes that what is true of the 12th Ward, is also true of the other Wards of the city.

That John Cavanagh, put down in the list as Warden in the 14th Ward, deponent finds to live with a brother at corner of Mott and Murray streets; which brother keeps a second-class liquor store at the last-named place.

That Patrick Burnes, the Assistant Warden of said Ward, is put down in the Directory for 1865, as a " Clerk," and is reported to be an active politician.

That James Lawrence, named in said list as Warden of the 5th Ward, as appears by the official Pay-rolls from the Mayor's Office, was paid up to February 1, 1865, and since that date (that is within the last 15 days), one Michael McManus has been made Health Warden in place of said Lawrence in said 5th Ward, and McManus is the keeper of a notorious liquor store and gin-mill at 67 West Broadway, which deponent has visited to personally ascertain its condition. It is the resort of the idle and intemperate of the Ward, and is reported by the police as a low gin-mill.

That in place of William Reynolds, reported in Valentine's Manuel for 1864, as Assistant Warden in the 21st Ward for that year, it appears by a report from the Mayor's Office, made this day, that one Bernard McCabe has been put in the place of said Reynolds. Said McCabe is reported in the City Directory for 1865, as keeping a liquor store at 391 3d Avenue in said city. And said McCabe is a well-known active politician.

And it is reported by the police, that during some portion of the past or present year, one John Murray has been Assistant Health Warden in said 21st Ward.

That, in the 22d Ward, one Lawrence Morrisey has, during the present year, been made Assistant Health Warden in the place of James A. McCormick, which latter has been made Warden in said Ward.

That said McCormick was known as a sporting man and politician in said Ward, and had no active business, as deponent is informed and believes. That the police report said Morrisey as a bar-tender, and deponent has reason to believe that, for years before his appointment, he was a bartender in a liquor and vitualling house much frequented by car-drivers, at the corner of 49th street and 8th Avenue in said city.

That Terence Foley, the Assistant Warden in the 6th Ward, keeps and personally attends a liquor store at 41 Elm street, near the Tombs of said city, which is a celebrated head-quarters not only of the politicians of said Ward but of the city officials generally.

Deponent further says, that it is generally understood, among those knowing the facts, that the appointments of Health and Assistant Health Wardens are, in a large measure, made to reward political service or to secure political influence, and from such reason there are many changes occurring during the year.

That it is the custom in the city that Health Officers, and other officers, should pay a portion of their salary to support the political expenses of their party, and such is now the fact.

That Vermilyea, the Assistant Health Warden in said 12th Ward, has generally been an Inspector of Primary Elections in said Ward, and also an Inspector on regular election days.

That the Health Wardens and Assistant Health Wardens, as is generally understood, are selected by the Aldermen, and required to be nominated to the Board of Aldermen, by the City Inspector, and are, for the most part, the well-known political agents and runners of the Aldermen of the respective Wards.

That the foregoing facts can be substantiated by the evidence of numerous persons.

JAMES MULLIGAN.

Subscribed and sworn to, before me,
this 15th day of February, 1865,

HENRY A. TAILER,

[STAMP, 5c.]   Com'r of Deeds.

www.ingramcontent.com/pod-product-compliance
Lightning Source LLC
Chambersburg PA
CBHW022015190326
41519CB00010B/1530